George Thomas Allen

Tables of Parabolic Curves

For the Use of Railway Engineers and Others

George Thomas Allen

Tables of Parabolic Curves
For the Use of Railway Engineers and Others

ISBN/EAN: 9783744678391

Printed in Europe, USA, Canada, Australia, Japan

Cover: Foto ©berggeist007 / pixelio.de

More available books at **www.hansebooks.com**

TABLES

OF

PARABOLIC CURVES

FOR THE USE OF RAILWAY ENGINEERS

AND OTHERS

BY

GEORGE T. ALLEN

ASSOC. M. INST. C.E.

London

E. & F. N. SPON, Limited, 125 STRAND

New York

SPON & CHAMBERLAIN, 12 CORTLANDT STREET

1898

[Entered at Stationers' Hall.]

PREFACE.

THE Tables of Parabolic Curves have been calculated by the author, and are now brought to the knowledge of engineers with a view of supplying a want: that is, the practical tracing of curves with a gradually decreasing and increasing radius, instead of the usual circular curves. By this means it is hoped that the great shocks caused by change of direction in such a mass as a moving train on entering a circular curve may be avoided, resulting, no doubt, in great saving as regards material in working, as well as much greater convenience to those who are travelling.

The use of the Tables is also strongly advised in laying out irrigation and navigable canals, as tending largely, in the former case, to diminish erosion, and in the latter to facilitate navigation, and also to prevent wash. Indeed, in all engineering works where a change in direction of moving masses is involved, these tables are recommended.

The Author is further of opinion that Parabolic Curves fit more easily into the natural surface of mountainous or undulating country than circular curves do, thus diminishing in a considerable degree the amount of cutting and embankment in new works.

The Tables are also useful in the design of structures where the contour is a parabola, the accurate length being easily obtainable.

GEORGE T. ALLEN.

GRAVESEND: 1898.

EXPLANATION.

THE tables give, as will be seen on inspection, the lengths of the curve, the tangent, the radius at apex, and the half-chord for the half-angle of intersection A B D or D B C (see Fig. 1), for every ten

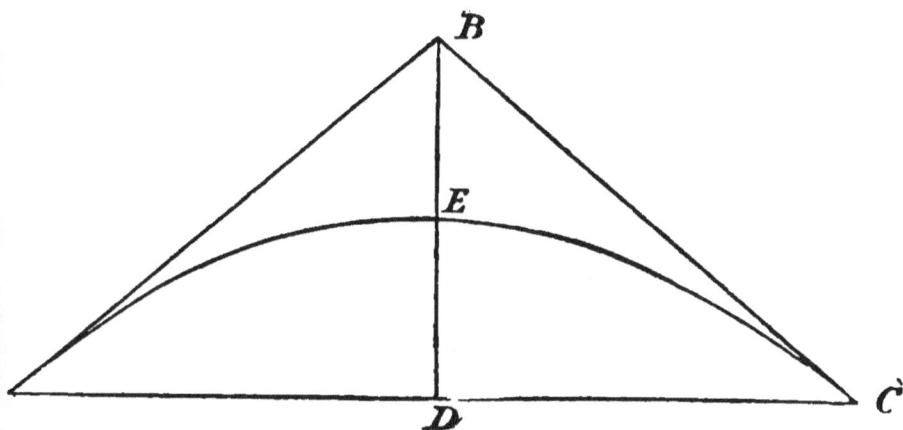

FIG. 1.

minutes of that angle, the height B D being unity. The bisectrix B E, and the height of the curve D E, are each equal to ·5, or the half of unity.

In setting out the curves practically, the bisectrix is usually fixed on the ground; then, to obtain the actual corresponding lengths of the curve, tangent, radius at apex, &c., it is necessary to multiply the lengths given in the tables by twice the length of the bisectrix as fixed upon.

If the length of the tangent is fixed upon and not that of the bisectrix, this length, divided by the quantity A B or B C in the columns, gives the total height from the chord to the intersection, and the true length of curve, radius at apex, or half-chord, will be found by multiplying each quantity by the height obtained.

As in all curve books, interpolation must be resorted to for the odd minutes when the half-angle is not an even ten, and the resulting numbers are to be multiplied by the double bisectrix, or height from chord to intersection as above.

The method of setting out the intermediary points in the curve is the following :—

The length of the bisectrix being determined upon, the tangent is divided into so many parts as it may be thought necessary to adopt of intermediary points, as at F, G, H, I, K, B (see Fig. 2), in which example the tangent is divided into six parts. The bisectrix is then divided by the square of this number, in this case by 36; the result is now multiplied by the squares of 1, 2, 3, &c., to 6—that is, by 1, 4, 9, &c., to 36—and these are the lengths of the

offsets F F,, G G,, H H,, &c., to B E, which is the bisectrix. It is an advantage to adopt for the length
of the bisectrix a quantity which is the square, or a
multiple of the square, of the number of divisions it
has been decided to divide the tangent into, as this

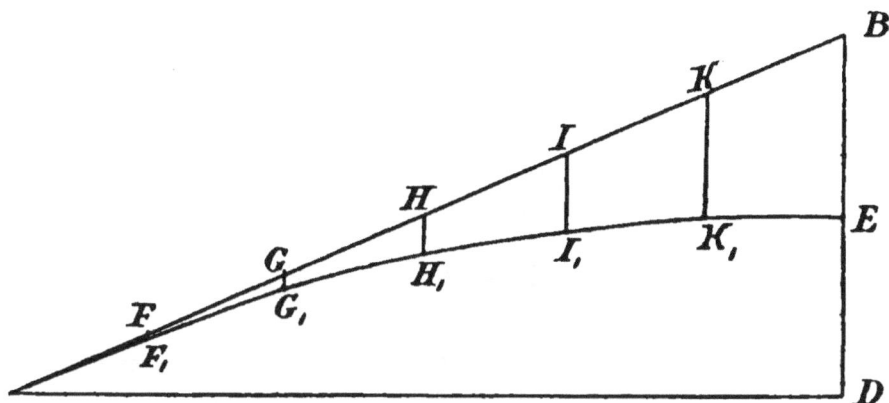

Fig. 2.

facilitates the calculations of the offsets, and is fre-
quently within the competency of the operator.

The offsets are in all cases parallel to the bisectrix.

Example I.—The angle of intersection of two
tangents equals 139°, and the bisectrix is equal to
5 chains; required the length of tangents, the length
of the curve, and the radius at apex.

Here, the half-angle of intersection A B D (Fig. 1)

is equal to 69° 30', and the total height is equal to 10 chains.

		Chains	Links
Tangent. . . $= 2 \cdot 855 \times 10$ =	28	55
Curve . . . $= 5 \cdot 471 \times 10$ =	54	71
Radius at apex $= 7 \cdot 153 \times 10$ =	71	53

Note.—Metres, yards, feet or any other standard of length is used in exactly the same manner.

EXAMPLE II.—The angle of intersection of two tangents equals 132° 40', and the bisectrix equals 4 ch. 90 lks. Required the length of the curve, tangent, radius at apex, and the offsets for 7 intermediate points.

The half-angle of intersection in this case is 66° 20', and the total height from chord to intersection 9 ch. 80 lks. Therefore,

		Chains	Links
Tangent . . $= 2 \cdot 491 \times 9 \cdot 80$ =	24	41
Curve . . . $= 4 \cdot 705 \times 9 \cdot 80$ =	46	10
Radius at apex $= 5 \cdot 206 \times 9 \cdot 80$ =	51	01

For 7 intermediate points the tangent is to be divided into that number of parts, and the bisectrix is to be divided by the square of this number or 49, and this gives the modulus for the offsets: that is,

$$ M = \frac{4 \cdot 90}{49} = 10 \text{ links.} $$

The offsets then are

			Chains	Links
10 ×	1	=		10
10 ×	4	=		40
10 ×	9	=		90
10 ×	16	=	1	60
10 ×	25	=	2	50 .
10 ×	36	=	3	60
10 ×	49	=	4	90

the last being the bisectrix.

EXAMPLE III.—Required the length of the curve, the length of the tangent, and the radius at apex for the insertion of a curve between two tangents, the angle of intersection being 145° 8', and the length of the bisectrix being 2 ch. 30 lks. Find also the offsets for 6 intermediate points.

The half-angle of intersection is 72° 34'.

The tangent for 72° 40' is 3·3564

    ,,   ,,   72° 30' ,, 3·3255

Difference . . . . . 0·0309

The tangent for 72·30 . 3·3255

Add $\frac{4}{10}$ of difference . . 0·01236

Gives tangent for 72° 34' 3·33786

which, multiplied by 4·60, gives 15 ch. 35 lks., the length of the tangent sought.

The curve for 72° 40' is 6·5106

„ „ 72° 30' „ 6·4467

Difference . . . . 0·0639

The curve for 72° 30' is 6·4467

Add $\frac{4}{10}$ of difference . 0·02556

Curve for 72° 34' . . 6·47226

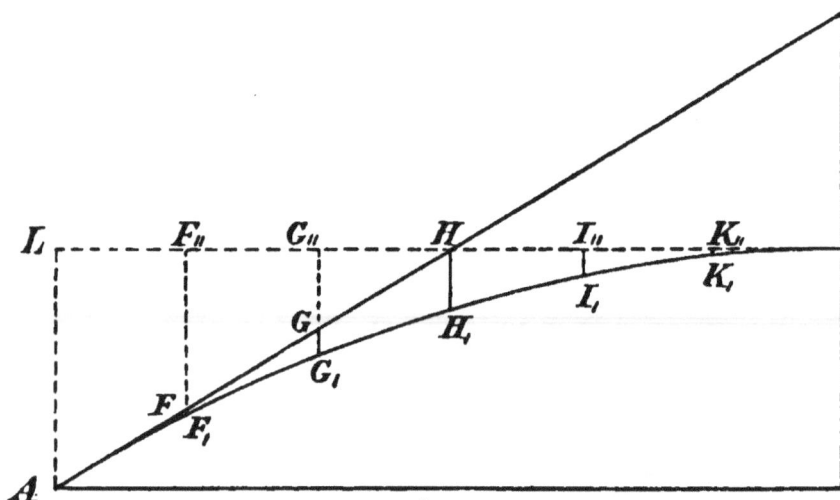

Fig. 3.

which, multiplied by 4·60, gives 29 ch. 77 lks., the length of the curve.

The radius at apex for 72° 40' is 10·2660

„ „ 72° 30' „ 10·0590

Difference . . . . 0·2070

Radius at apex for 72° 30′ is 10·0590
Add $\frac{4}{10}$ of difference . . . 0·08280

Radius at apex for 72° 34′ . 10·14180

which, multiplied by 4·60, gives 46 ch. 65 lks.

The bisectrix being equal to 230 links, and the intermediate points to 6,

$$M = \frac{230}{36} = 6\cdot388 \text{ links.}$$

The offsets are, therefore,

				Chains	Links
6·388	×	1	=		6·388
6·388	×	4	=		25·552
6·388	×	9	=		57·492
6·388	×	16	=	1	2·208
6·388	×	25	=	1	59·700
6·388	×	36	=	2	29·968

or 2 ch. 30 lks., which is the bisectrix.

The number of decimals in these results is decided by the degree of accuracy required by the operator.

At times it may be found convenient, as in the case of large curves, to set off from a line parallel to the chord and passing through the apex. In this case the half-chord is divided into a convenient number of parts in the manner in which it has been recommended to divide the tangent. These divisions are set off along E L (Fig. 3) at K‚ I‚ H G‚‚‚ &c., and the same ordinates are used which are calcu-

lated for setting out from the tangent. K K$_{,,}$ is equal
to F F$_{,}$. I$_{,}$ I$_{,,}$ is equal to G G$_{,}$, or half the curve may

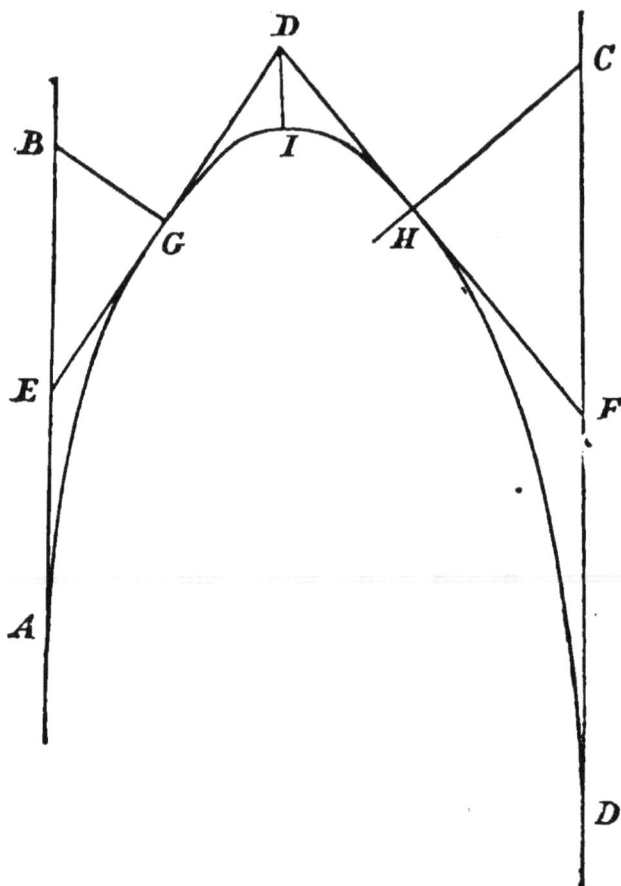

Fig. 4.

be traced from the half-chord, and the other half
from the tangent.

It happens occasionally in practice that it is necessary to trace a curve joining two tangents which are possibly parallel, or even whose angle of intersection is negative, and it would appear at first sight that it would be impossible to effect the desired purpose by employing the parabola, the nature of this curve requiring divergent tangents. As an example of this case see Fig. 4, in which A B and C D are the proposed tangents. By adopting a

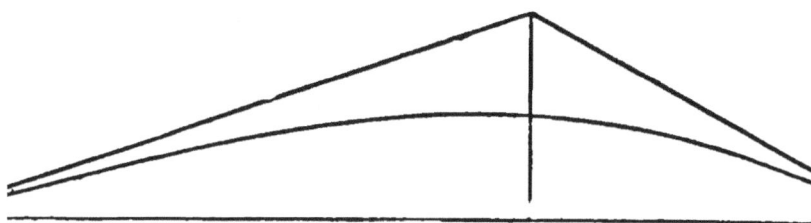

FIG. 5.

point D, and secondary tangents D E and D F (such lines are usually given as portions of the survey), and these lines being divided respectively in the points G and H, making D G and D H equal, the half parabola A G is traced from the tangent A B and angle E B G, and similarly the half-parabola D H, and finally the whole parabola G I H is inserted on the tangents G D D H within the angle G D H.

The radius of curvature at the apex of the curve

(the point E iu Fig. 1) is given in the column under that heading, the height from chord to intersection being unity. If, however, it is necessary to ascertain the radius at the tangent points, the following formula applies:

$$\rho = \frac{\overline{A\,B^3}}{\overline{A\,D}}.$$

The radius at any other point in the curve can be obtained by calculating A B and A D on the assumption that the point is a tangent point.

It may be stated that each half of the curve is quite independent of the other, as in the case of

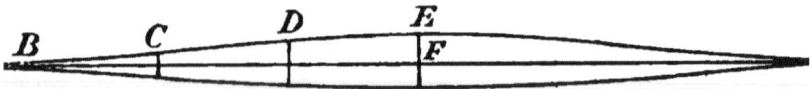

Fig. 5, where two different curves are tangent to each other, the only condition being that the height must be the same in each case.

The outer elevation and inner depression of the rails in a curve can be set to parabolic ordinates by making the elevation or depression at B one-eighth of the total rise E F, and at D seven-eighths of the same quantity (see Fig. 6). In a long curve it is

advisable to adopt the points calculated for setting out the curve as the ones where correction for elevation or depression is effected.

To set out a tangent at any point of the curve (see Fig. 3), the following is the proceeding : If a tangent is required passing through $F_{,}$ set off on A L, commencing at A, a quantity equal to the offset $F F_{,,}$ and this point is in the tangent ; similarly, if a tangent is required at $G_{,}$ set off on A L a quantity equal to the offset $G G_{,,}$ and so on, for any point. The same result is obtained by setting out lengths along E B which are severally equal to K, $K_{,,,}$, I, $I_{,,,}$, &c. For a point falling between any of the points considered above, it is necessary to know its offset from the tangent, and to proceed exactly in the same manner. The offsets must in all cases be parallel to the bisectrix.

PRÉFACE.

Ces Tables de Courbes Paraboliques ont été calculées par l'auteur et sont aujourd'hui portées à la connaissance des ingénieurs avec le but de remplir un besoin : c'est à dire, pour fournir une méthode pratique de tracer une courbe dont le rayon décroît ou croît graduellement au lieu de se servir de la courbe circulaire. Par ce moyen on espère que les grandes secousses que l'on éprouve par les changements de direction d'une grande masse comme un train en mouvement quand ceci entre dans une courbe circulaire seront évitées ; ce qui résulte en grande épargne de matériel roulant et fixe, ainsi que la plus grande convenance des voyageurs.

L'usage de ces tables est aussi fortement raccommandé dans le tracé des canaux d'irrigation et de navigation : dans le premier cas pour la diminution de l'érosion des bords, et dans le second pour faciliter la navigation et aussi pour empêcher les dédommagements des vagues causées par le passage des bateaux. En général, quand il est question de

c

changement de direction de grandes masses, ces tables sont fort propres à employer.

L'auteur est aussi d'opinion, que les courbes paraboliques se raccordent beaucoup plus facilement à la surface naturelle du terrain dans les pays montagneux ou ondulants que n'est le cas des courbes circulaires ; ce qui diminue dans un degré appréciable la quantité de remblais et déblais nécessaires sur les nouveaux travaux.

Les tables sont aussi d'utilité dans les projets de construction, là où le contour est une parabole, car on en obtient la longueur exacte de la courbe avec grande facilité.

GEORGE T. ALLEN,

Ingénieur.

GRAVESEND : 1898.

EXPLICATION.

Ces tables donnent la longueur de la courbe parabolique, ainsi que celle de la tangente, le rayon au
sommet de la courbe, et la demi-corde pour la moitié
de l'angle compris entre les tangentes ; c'est à dire :

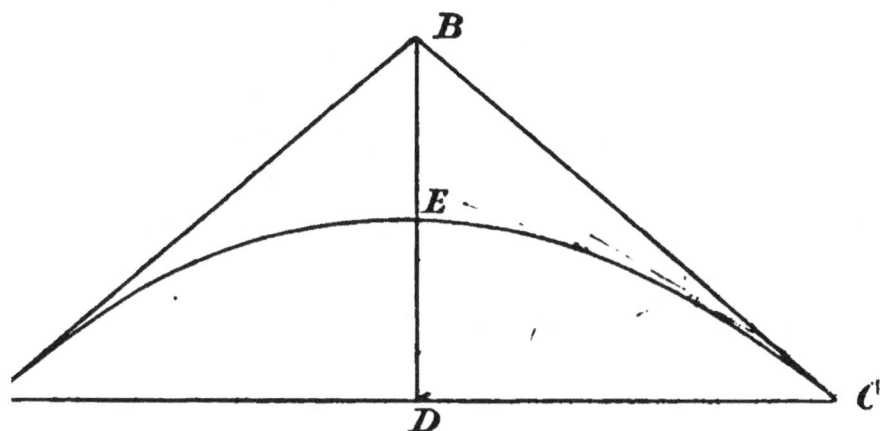

Fig. 1.

pour l'angle A B D ou l'angle D B C (voir la figure
No. 1), pour chaque dix minutes d'ouverture ; la
hauteur B D est en tous cas égale à l'unité. La
bissectrice B E et la hauteur de la courbe D E
équivalent 0·5 ou bien la moitié de l'unité

c 2

Dans la pratique quand on trace la courbe, la longueur de la bissectrice est ordinairement fixée sur le terrain ; alors, pour obtenir la vraie longueur de la courbe, la tangente, le rayon au sommet, on multiplie les chiffres données dans les colonnes par le double de la longueur de la bissectrice qui a été fixée.

Si au lieu de la bissectrice on fixe la longueur de la tangente, on divise cette longueur par la quantité donnée dans la colonne des tangentes A B ou B C, et le numéro qui résulte est la hauteur totale, ou la bissectrice doublée ; et pour avoir la vraie longueur de la courbe, le rayon au sommet ou la demi-corde on doit multiplier le numéro obtenu par le nombre dans les colonnes.

On doit se servir de l'interpolation pour les angles qui ont des minutes entre les dix.

La méthode de tracer les points intermédiaires, c'est comme suit.

Quand on a décidé de la longueur de la bissectrice, on divise la tangente en autant de parties égales que l'on veut adopter de points intermédiaires, comme on voit sur la figure No. 2, F, G, H, I, K, B, dans quel cas la tangente est divisée en six parties. Alors on divise la bissectrice par le carré de ce nombre, c'est à dire, par 36. Ce résultat est multiplié par les carrés de 1, 2, 3, etc., jusqu'à 6, c'est à dire, par 1, 4, 9, etc., jusqu'à 36, et ces résultats sont effectivement les longueurs des ordonnés F F, G G, H H, et cetera jusqu'à B E que l'on voit est la bissectrice. On peut remarquer qu'on trouve avautage de fixer pour la longueur de la bissectrice une quantité qui est le

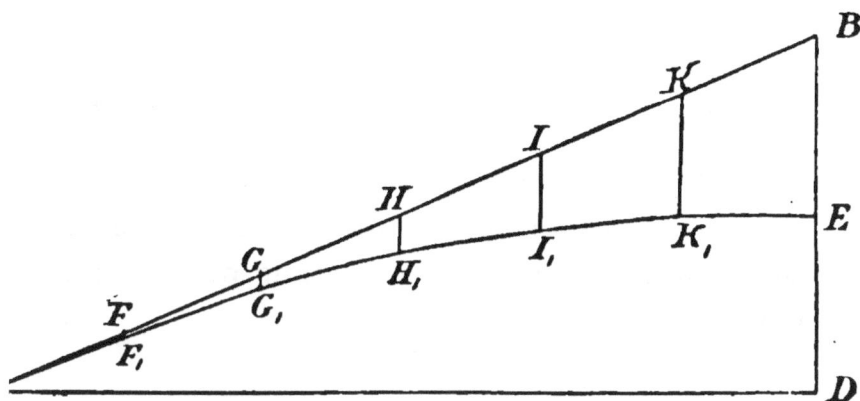

Fig. 2.

carré ou un multiple du carré du nombre dans lequel on a décidé de diviser la tangente. La raison, c'est qu'ainsi on facilite le calcul des ordonnés, et souvent cette démarche se trouve à la compétence de l'opérateur. Ces ordonnés sont toujours parallèles à la bissectrice.

EXEMPLE Ier. L'angle de l'intersection de deux tangentes equivaut 139°, et la bissectrice a 50 mètres de longueur. On demande la longueur de la tangente, celle de la courbe et le rayon au sommet (à l'apex).

Dans ce cas, la moitié de l'angle de l'intersection A B D (Fig. 1) vaut 69° 30', et la hauteur totale est de 100 mètres ; donc

La tangente . . = 2·8554 × 100 = 285·54 ^{m.}
La courbe . . . = 5·4713 × 100 = 547·13
Le rayon au sommet = 7·1536 × 100 = 715·36

qui est le résultat demandé.

Nota.—On peut employer de la même façon une base de longueur quelconque, comme pieds anglais ou autres.

EXEMPLE II. L'angle d'intersection de deux tangentes vaut 132° 40' et la bissectrice est de 49 mètres. On demande la longueur de la courbe, la tangente, le rayon au sommet et les ordonnées pour 7 points intermédiaires.

La moitié de l'angle d'intersection dans ce cas est 66° 20', et la hauteur totale de la corde à l'intersection 98 mètres. Alors

La tangente . . = 2·4911 × 98 = 244·12 ^{m.}
La courbe . . . = 4·7054 × 98 = 461·12
Le rayon au sommet = 5·2060 × 98 = 510·18

Pour 7 points intermédiaires on doit diviser la tangente en 7 parties égales et la bissectrice doit se diviser par 49, ce qui donne un module pour les ordonnées; c'est à dire :

$$M = \frac{49}{49} = 1 \text{ mètre.}$$

Les ordonnées sont alors

$$
\begin{array}{rcl}
 & & \text{m.} \\
1 \times 1 & = & 1 \\
1 \times 4 & = & 4 \\
1 \times 9 & = & 9 \\
1 \times 16 & = & 16 \\
1 \times 25 & = & 25 \\
1 \times 36 & = & 36 \\
1 \times 49 & = & 49
\end{array}
$$

Ce dernier on reconnaît pour la bissectrice.

EXEMPLE III. On demande la longueur de la courbe de la tangente et aussi le rayon de courbure au sommet pour conjoindre les deux tangentes dont l'angle de l'intersection est 145° 8' et la longueur de la bissectrice est 23 mètres. En plus, il faut les ordonnés pour six points intermédiaires.

La moitié de l'angle vient à 72° 34'.

La tangente pour 72° 40' est
(voir les tables) } 3·3564

    „   „   72° 30' est   3·3255

Différence . . . . 0·0309

La tangente pour 72° 30′ est   3·3255

Ajoutez $\frac{4}{10}$ de la différence pour les 4 minutes . . . . . } 0·01236

Donc la tangente pour 72° 34′ est   3·33786

En multipliant ce chiffre par la hauteur totale 46 mètres on trouve la longueur de la tangente, 152·541 mètres.

La courbe pour 72° 40′ est . . 6·5106

   „   „   72° 30′ est . . 6·4467

Différence . . . . 0·0639

La courbe pour 72° 30′ est .. 6·4467

Ajoutez $\frac{4}{10}$ de la différence .. 0·02556

Donc la courbe pour 72° 34′ est   6·47226

En multipliant ce nombre par 46 mètres on trouve la longueur de la courbe, 297·723 mètres.

Le rayon au sommet pour 72° 40′ est . . . . } 10·2660

   „   „   72° 30′ est 10·0590

Différence . . . . 0·2070

Le rayon au sommet pour 72° 30′ est . . . . } 10·0590

Ajoutez $\frac{4}{10}$ de la différence . . 0·08280

Donc le rayon au sommet pour 72° 34′ est . . . . . . } 10·14180

Ce nombre multiplié par 46 mètres donne 466·522 mètres, le rayon cherché.

Pour obtenir les ordonnés intermédiaires, comme la bissectrice equivaut 23 mètres et le nombre d'ordonnés 6, le module

$$M = \frac{23}{36} = 0 \cdot 638 \text{ mètre.}$$

Les ordonnés sont alors

$$
\begin{aligned}
&\ \overset{\text{m.}}{}\\
0 \cdot 638 \times 1 &= 0 \cdot 63\\
0 \cdot 638 \times 4 &= 2 \cdot 55\\
0 \cdot 638 \times 9 &= 5 \cdot 74\\
0 \cdot 638 \times 16 &= 10 \cdot 20\\
0 \cdot 638 \times 25 &= 15 \cdot 95\\
0 \cdot 638 \times 36 &= 22 \cdot 96
\end{aligned}
$$

c'est à dire 23 mètres, qui est la bissectrice.

Le nombre des points décimales dépend du degré d'approximation voulu.

Quelquefois, surtout quand il s'agit de grandes courbes, on facilite les opérations en faisant la tracée sur une ligne parallèle à la corde qui en même temps passe par le sommet de la courbe. Voir la figure 3. Dans ce cas la ligne E L est divisée dans un nombre de parties égales de la même façon que l'on a fait pour la tangente, et les mêmes ordonnés sont calculés comme si l'on voulait tracer sur la tangente, mais on

les trace en positions inverses. De fait, K, K₎ est
égal à F F₎, et I, I₎ est égal à G G₎, et ainsi de suite.
On peut tracer la moitié de la courbe de la tangente
et l'autre moitié de la ligne E L à volonté.

Dans la pratique il arrive qu'il est nécessaire de
tracer une courbe entre deux tangentes sensiblement

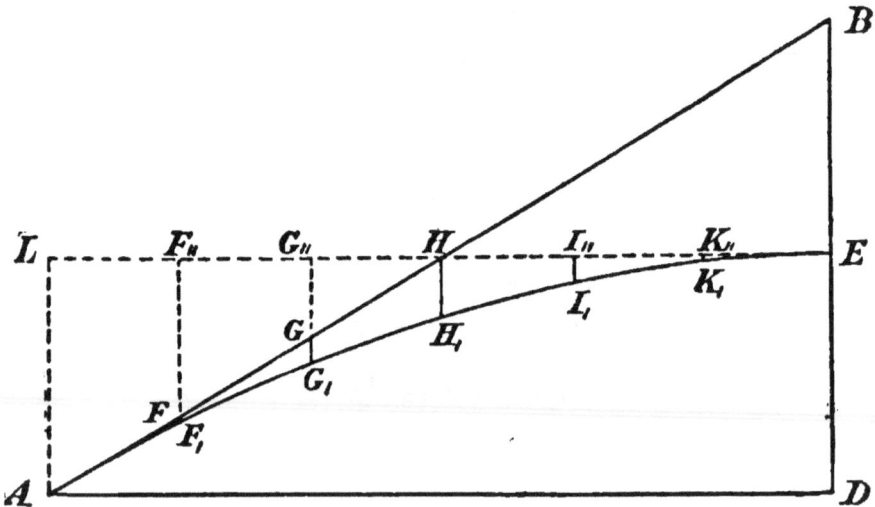

Fɪɢ. 3.

parallèles ou dont l'angle d'intersection est même
négative, et à première vue on le croirait impossible,
parce que la nature de la parabole demande des tangentes divergentes. Comme exemple de ce cas, voir
la figure No. 4 où A B et C D sont les tangentes dont
il s'agit. On interpose un point D formant des
tangentes secondaires D E et D F (les lignes telles

ordinairement forment partie de l'arpentage). Alors
on divise ces lignes respectivement en G et H et de

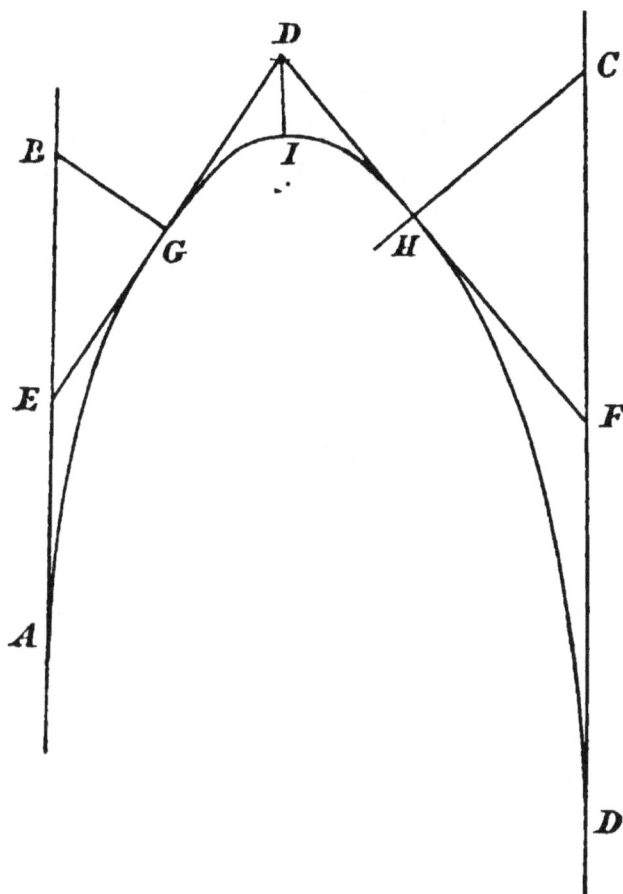

Fig. 4.

façon que D G = D H. La demi-parabole A G est
tracée de la tangente A B avec l'angle E B G, et égale-

ment la demi-parabole D H ; et finalement la parabole
G'I H est tracée entre les tangentes G D, D H dans
l'angle G D H.

Le rayon de courbure au sommet de la courbe
(E, Fig. 1) est donné dans les colonnes, la hauteur
totale de la corde à l'intersection étant toujours
l'unité. Si l'on a besoin de connaître le rayon de
courbure aux points de tangence on a

$$\rho = \frac{\overline{A\ B^3}}{\overline{A\ D}}.$$

Fig. 5.

On peut remarquer que chaque moitié de la courbe
est parfaitement indépendante de l'autre, comme dans
la figure No. 5 où les deux courbes sont tangentes
l'une à l'autre; la seule condition, c'est qu'elles ont la
même hauteur.

Dans une courbe de chemin de fer, l'élévation du
rail extérieur et la dépression du rail intérieur peu-
vent s'accorder à la forme parabolique en faisant la

hausse à B (voir Fig. 6) de la valeur d'une huitième
de la hausse totale E F, et la quantité à D de sept-
huitièmes de la même. Dans une grande courbe il
est mieux de calculer ces corrections pour les points
qui correspondent aux ordonnés de la tracée.

Pour tracer une tangente à un point quelconque
de la courbe (voir Fig. 3) on fait comme suit. Si
on demande une tangente qui traverse le point F,
on trace sur A L depuis le point A une quantité
égale à F F,; le point résultant est dans la tangente

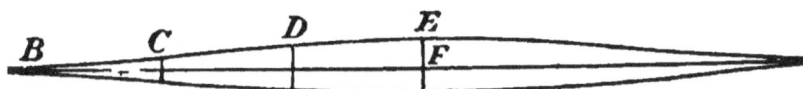

FIG. 6.

voulue. Aussi, si l'on a besoin d'une tangente à G,
il faut tracer sur la même ligne la quantité G G,, et
ainsi de suite pour tout autre point. On obtient le
même résultat en traçant sur E B des distances égales
à K, K,,, I, I,, pour les points K, I, et cetera. Pour un
point qui se trouve entre deux des points déjà con-
sidérés, il faut calculer l'ordonné exacte et poursuivre
la même construction. Les ordonnés sont en tous
les cas parallèles à la bissectrice.

TABLES

OF

PARABOLIC CURVES.

Angle A B D or D B C.	Curve A E C.	Tangent A B or B C.	Radius at Apex.	Half-Chord A D or D C.
40°	2·017018	1·305407	·7041	·839100
10	2·025452	1·308607	·7125	·844069
20	2·033939	1·311833	·7209	·849062
30	2·042481	1·315087	·7295	·854081
40	2·051078	1·318368	·7381	·859124
50	2·059730	1·321676	·7468	·864193
41°	2·068437	1·325013	·7557	·869287
10	2·077200	1·328378	·7646	·874407
20	2·086020	1·331771	·7736	·879553
30	2·094897	1·335192	·7827	·884725
40	2·103832	1·338643	·7920	·889924
50	2·112825	1·342123	·8013	·895151

Angle A B D or D B C.	Curve A E C.	Tangent A B or B C.	Radius at Apex.	Half-Chord A D or D C.
42°	2·121877	1·345633	·8107	·900404
10	2·130989	1·349172	·8203	·905685
20	2·140161	1·352742	·8299	·910994
30	2·149394	1·356342	·8397	·916331
40	2·158687	1·359972	·8495	·921697
50	2·168042	1·363634	·8595	·927091
43°	2·177460	1·367327	·8696	·932515
10	2·186941	1·371052	·8798	·937968
20	2·196486	1·374809	·8901	·943451
30	2·206095	1·378598	·9005	·948965
40	2·215769	1·382420	·9111	·954508
50	2·225509	1·386275	·9218	·960083
44°	2·235315	1·390164	·9326	·965689
10	2·245189	1·394086	·9435	·971326
20	2·255130	1·398042	·9545	·976996
30	2·265140	1·402032	·9657	·982697
40	2·275219	1·406057	·9770	·988432
50	2·285368	1·410118	·9884	·994199

Angle A B D or D B C.	Curve A E C.	Tangent A B or B C.	Radius at Apex.	Half-Chord A D or D C.
45°	2·295587	1·414214	1·0000	1·000000
10	2·305878	1·418345	1·0117	1·005835
20	2·316242	1·422513	1·0235	1·011704
30	2·326678	1·426718	1·0355	1·017607
40	2·337188	1·430960	1·0476	1·023546
50	2·347773	1·435239	1·0599	1·029520
46°	2·358433	1·439556	1·0723	1·035530
10	2·369169	1·443912	1·0849	1·041577
20	2·379982	1·448306	1·0976	1·047660
30	2·390874	1·452740	1·1105	1·053780
40	2·401845	1·457213	1·1235	1·059938
50	2·412895	1·461726	1·1366	1·066134
47°	2·424025	1·466279	1·1500	1·072369
10	2·435237	1·470874	1·1635	1·078642
20	2·446531	1·475509	1·1771	1·084955
30	2·457909	1·480187	1·1910	1·091308
40	2·469371	1·484907	1·2049	1·097702
50	2·480919	1·489670	1·2191	1·104136

D

Angle A B D or D B C.	Curve A E C.	Tangent A B or B C.	Radius at Apex.	Half-Chord A D or D C.
48°	2·492553	1·494476	1·2335	1·110612
10	2·504274	1·499327	1·2480	1·117130
20	2·516083	1·504221	1·2627	1·123691
30	2·527981	1·509160	1·2776	1·130294
40	2·539969	1·514145	1·2926	1·136941
50	2·552049	1·519176	1·3079	1·143633
49°	2·564222	1·524253	1·3233	1·150368
10	2·576489	1·529377	1·3390	1·157149
20	2·588850	1·534549	1·3548	1·163976
30	2·601307	1·539769	1·3709	1·170850
40	2·613862	1·545038	1·3871	1·177770
50	2·626515	1·550356	1·4036	1·184738
50°	2·639267	1·555724	1·4203	1·191754
10	2·652120	1·561142	1·4372	1·198818
20	2·665074	1·566612	1·4543	1·205933
30	2·678131	1·572134	1·4716	1·213097
40	2·691293	1·577708	1·4892	1·220312
50	2·704561	1·583335	1·5070	1·227579

Angle A B D or D B C.	Curve A E C.	Tangent A B or B C.	Radius at Apex.	Half-Chord A D or D C.
51°	2·717937	1·589016	1·5250	1·234897
10	2·731421	1·594751	1·5432	1·242268
20	2·745014	1·600542	1·5617	1·249693
30	2·758718	1·606388	1·5804	1·257172
40	2·772535	1·612291	1·5995	1·264706
50	2·786467	1·618251	1·6187	1·272296
52°	2·800514	1·624269	1·6383	1·279942
10	2·814678	1·630346	1·6580	1·287645
20	2·828960	1·636483	1·6781	1·295406
30	2·843363	1·642680	1·6984	1·303225
40	2·857888	1·648938	1·7190	1·311105
50	2·872535	1·655257	1·7399	1·319044
53°	2·887308	1·661640	1·7610	1·327045
10	2·902208	1·668086	1·7825	1·335107
20	2·917236	1·674597	1·8043	1·343233
30	2·932394	1·681173	1·8263	1·351422
40	2·947684	1·687815	1·8487	1·359676
50	2·963108	1·694524	1·8714	1·367996

Angle A B D or D B C.	Curve A E C.	Tangent A B or B C.	Radius at Apex.	Half-Chord A D or D C.
54°	2·978667	1·701302	1·8944	1·376382
10	2·994364	1·708148	1·9178	1·384835
20	3·010200	1·715064	1·9414	1·393357
30	3·026177	1·722051	1·9654	1·401948
40	3·042298	1·729110	1·9898	1·410610
50	3·058564	1·736241	2·0145	1·419343
55°	3·074977	1·743447	2·0396	1·428148
10	3·091539	1·750727	2·0650	1·437027
20	3·108253	1·758084	2·0909	1·445980
30	3·125121	1·765517	2·1171	1·455009
40	3·142144	1·773029	2·1436	1·464115
50	3·159326	1·780620	2·1706	1·473298
56°	3·176668	1·788292	2·1980	1·482561
10	3·194173	1·796045	2·2258	1·491904
20	3·211843	1·803881	2·2540	1·501328
30	3·229681	1·811801	2·2826	1·510835
40	3·247689	1·819806	2·3117	1·520426
50	3·265870	1·827898	2·3412	1·530102

Angle A B D or D B C.	Curve A E C.	Tangent A B or B C.	Radius at Apex.	Half-Chord A D or D C.
57°	3·284226	1·836078	2·3712	1·539865
10	3·302760	1·844348	2·4016	1·549715
20	3·321474	1·852707	2·4325	1·559655
30	3·340372	1·861159	2·4639	1·569686
40	3·359457	1·869704	2·4958	1·579808
50	3·378731	1·878344	2·5282	1·590024
58°	3·398197	1·887080	2·5611	1·600334
10	3·417857	1·895914	2·5945	1·610742
20	3·437716	1·904847	2·6284	1·621247
30	3·457775	1·913881	2·6629	1·631852
40	3·478039	1·923017	2·6980	1·642558
50	3·498511	1·932258	2·7336	1·653366
59°	3·519194	1·941604	2·7698	1·664279
10	3·540091	1·951058	2·8066	1·675299
20	3·561206	1·960621	2·8440	1·686426
30	3·582543	1·970294	2·8821	1·697663
40	3·604105	1·980081	2·9207	1·709012
50	3·625896	1·989982	2·9600	1·720474

Angle A B D or D B C.	Curve A E C.	Tangent A B or B C.	Radius at Apex.	Half-Chord A D or D C.
60°	3·647919	2·000000	3·0000	1·732051
10	3·670179	2·010136	3·0406	1·743745
20	3·692679	2·020393	3·0819	1·755559
30	3·715423	2·030772	3·1240	1·767494
40	3·738416	2·041276	3·1668	1·779552
50	3·761662	2·051906	3·2103	1·791736
61°	3·785165	2·062665	3·2546	1·804048
10	3·808930	2·073556	3·2996	1·816489
20	3·832961	2·084579	3·3455	1·829063
30	3·857263	2·095738	3·3921	1·841771
40	3·881841	2·107036	3·4396	1·854616
50	3·906699	2·118474	3·4879	1·867600
62°	3·931843	2·130054	3·5371	1·880726
10	3·957277	2·141781	3·5872	1·893997
20	3·983007	2·153655	3·6382	1·907415
30	4·009039	2·165681	3·6902	1·920982
40	4·035377	2·177859	3·7431	1·934702
50	4·062027	2·190195	3·7970	1·948577

Angle A B D or D B C.	Curve A E C.	Tangent A B or B C.	Radius at Apex.	Half-Chord A D or D C.
63°	4·088995	2·202689	3·8519	1·962610
10	4·116287	2·215346	3·9078	1·976805
20	4·143910	2·228168	3·9647	1·991164
30	4·171869	2·241158	4·0228	2·005690
40	4·200170	2·254320	4·0820	2·020386
50	4·228821	2·267657	4·1423	2·035256
64°	4·257827	2·281172	4·2037	2·050304
10	4·287196	2·294868	4·2664	2·065532
20	4·316935	2·308750	4·3303	2·080944
30	4·347051	2·322820	4·3955	2·096544
40	4·377551	2·337083	4·4620	2·112335
50	4·408444	2·351542	4·5297	2·128321
65°	4·439737	2·366202	4·5989	2·144507
10	4·471438	2·381065	4·6695	2·160896
20	4·503555	2·396137	4·7415	2·177492
30	4·536095	2·411421	4·8149	2·194300
40	4·569069	2·426922	4·8899	2·211323
50	4·602486	2·442645	4·9665	2·228568

Angle A B D or D B C.	Curve A E C.	Tangent A B or B C.	Radius at Apex.	Half-Chord A B or D C.
66°	4·636355	2·458593	5·0447	2·246037
10	4·670686	2·474773	5·1245	2·263736
20	4·705487	2·491187	5·2060	2·281669
30	4·740768	2·507843	5·2893	2·299842
40	4·776540	2·524744	5·3743	2·318261
50	4·812813	2·541896	5·4612	2·336929
67°	4·849599	2·559305	5·5500	2·355852
10	4·886909	2·576975	5·6408	2·375037
20	4·924754	2·594914	5·7336	2·394489
30	4·963145	2·613126	5·8284	2·414214
40	5·002097	2·631618	5·9254	2·434217
50	5·041622	2·650396	6·0246	2·454506
68°	5·081731	2·669467	6·1260	2·475087
10	5·122437	2·688837	6·2298	2·495966
20	5·163756	2·708514	6·3360	2·517151
30	5·205701	2·728504	6·4447	2·538648
40	5·248287	2·748814	6·5560	2·560465
50	5·291529	2·769453	6·6699	2·582609

Angle A B D or D B C.	Curve A E C.	Tangent A B or B C.	Radius at Apex.	Half-Chord A D or D C.
69°	5·335443	2·790428	6·7865	2·605089
10	5·380046	2·811747	6·9059	2·627912
20	5·425353	2·833418	7·0282	2·651087
30	5·471381	2·855451	7·1536	2·674621
40	5·518148	2·877853	7·2820	2·698525
50	5·565674	2·900635	7·4137	2·722808
70°	5·61398	2·92380	7·5487	2·74748
10	5·66307	2·94737	7·6870	2·77254
20	5·71299	2·97135	7·8289	2·79802
30	5·76374	2·99574	7·9745	2·82391
40	5·81535	3·02057	8·1238	2·85023
50	5·86784	3·04584	8·2771	2·87700
71°	5·92124	3·07155	8·4344	2·90421
10	5·97556	3·09774	8·5960	2·93189
20	6·03084	3·12440	8·7619	2·96004
30	6·08710	3·15155	8·9322	2·98868
40	6·14436	3·17920	9·1073	3·01783
50	6·20266	3·20737	9·2872	3·04749

Angle A B D or D B C.	Curve A E C.	Tangent A B or B C.	Radius at Apex.	Half-Chord A D or D C.
72°	6·26202	3·23607	9·4721	3·07768
10	6·32247	3·26531	9·6623	3·10842
20	6·38405	3·29512	9·8578	3·13972
30	6·44678	3·32551	10·0590	3·17159
40	6·51069	3·35649	10·2660	3·20406
50	6·57583	3·38808	10·4791	3·23714
73°	6·64223	3·42030	10·6985	3·27085
10	6·70993	3·45317	10·9244	3·30521
20	6·77896	3·48671	11·1571	3·34023
30	6·84937	3·52094	11·3970	3·37594
40	6·92119	3·55587	11·6442	3·41236
50	6·99447	3·59154	11·8991	3·44951
74°	7·06926	3·62796	12·1620	3·48741
10	7·14561	3·66515	12·4333	3·52609
20	7·22356	3·70315	12·7133	3·56557
30	7·30317	3·74198	13·0024	3·60588
40	7·38449	3·78166	13·3010	3·64705
50	7·46757	3·82223	13·6094	3·68909

Angle A B D or D B C.	Curve A E C.	Tangent A B or B C.	Radius at Apex.	Half-Chord A D or D C.
75°	7·55248	3·86370	13·9282	3·73205
10	7·63928	3·90612	14·2578	3·77595
20	7·72802	3·94952	14·5987	3·82083
30	7·81878	3·99393	14·9515	3·86671
40	7·91164	4·03938	15·3166	3·91364
50	8·00666	4·08591	15·6947	3·96165
76°	8·10392	4·13357	16·0864	4·01078
10	8·20349	4·18238	16·4923	4·06107
20	8·30547	4·23239	16·9131	4·11256
30	8·40994	4·28366	17·3497	4·16530
40	8·51701	4·33621	17·8028	4·21933
50	8·62676	4·39012	18·2731	4·27471
77°	8·73931	4·44541	18·7617	4·33148
10	8·85475	4·50216	19·2694	4·38969
20	8·97320	4·56041	19·7973	4·44942
30	9·09478	4·62023	20·3465	4·51071
40	9·21963	4·68167	20·9181	4·57363
50	9·34786	4·74482	21·5133	4·63825

Angle A B D or D B C.	Curve A E C.	Tangent A B or B C.	Radius at Apex.	Half-Chord A D or D C.
78°	9·47964	4·80973	22·1335	4 70463
10	9·61511	4·87649	22·7802	4·77286
20	9·75441	4·94517	23·4547	4·84300
30	9·89772	5·01585	24·1588	4·91516
40	10·04522	5·08863	24·8941	4·98940
50	10·19709	5·16359	25·6627	5·06584
79°	10·35354	5·24084	26·4664	5·14455
10	10·51477	5·32049	27·3076	5·22566
20	10·68101	5·40263	28·1884	5·30928
30	10·85250	5·48740	29·1116	5·39552
40	11·02949	5·57493	30·0798	5·48451
50	11·21225	5·66533	31·0961	5·57638
80°	11·40107	5·75877	32·1634	5·67128
10	11·59626	5·85539	33·2856	5·76937
20	11·79814	5·95536	34·4663	5·87080
30	12·00707	6·05886	35·7098	5·97576
40	12·22343	6·16607	37·0204	6·08444
50	12·44763	6·27719	38·4032	6·19703

Angle A B D or D B C.	Curve A E C.	Tangent A B or B C.	Radius at Apex.	Half-Chord A D or D C.
81°	12·68010	6·39245	39·8635	6·31375
10	12·92131	6·51208	41·4072	6·43484
20	13·17175	6·63633	43·0409	6·56055
30	13·43196	6·76547	44·7716	6·69116
40	13·70255	6·89979	46·6072	6·82694
50	13·98415	7·03962	48·5563	6·96823
82°	14·27745	7·18530	50·6285	7·11537
10	14·58318	7·33719	52·8344	7·26873
20	14·90216	7·49571	55·1857	7·42871
30	15·23528	7·66130	57·6955	7·59575
40	15·58350	7·83443	60·3783	7·77035
50	15·94787	8·01564	63·2506	7·95302
83°	16·32954	8·20551	66·3304	8·14435
10	16·7298	8·4047	69·6383	8·3450
20	17·1500	8·6138	73·1974	8·5555
30	17·5917	8·8337	77·0338	8·7769
40	18·0566	9·0651	81·1770	9·0098
50	18·5466	9·3092	85·6606	9·2553

Angle A B D or D B C.	Curve A E C.	Tangent A B or B C.	Radius at Apex.	Half-Chord A D or D C.
84°	19·0638	9·5668	90·5231	9·5144
10	19·6104	9·8391	95·8083	9·7882
20	20·1891	10·1275	101·5667	10·0780
30	20·8029	10·4334	107·8565	10·3854
40	21·4549	10·7585	114·7450	10·7119
50	22·1490	11·1045	122·3110	11·0594
85°	22·8892	11·4737	130·646	11·4301
10	23·6805	11·8684	139·858	11·8262
20	24·5282	12·2913	150·075	12·2505
30	25·4386	12·7455	161·448	12·7062
40	26·4190	13·2347	174·158	13·1969
50	27·4776	13·7631	188·423	13·7267
86°	28·6246	14·3356	204·509	14·3007
10	29·8711	14·9579	222·738	14·9244
20	31·2309	15··6368	243·509	15·6048
30	32·7201	16·3804	267·318	16·3499
40	34·3582	17·1984	294·786	17·1693
50	36·1684	18·1026	326·705	18·0750

Angle A B D or D B C.	Curve A E C.	Tangent A B or B C.	Radius at Apex.	Half-Chord A D or D C.
87°	38·1799	19·1073	364·090	19·0811
10	40·428	20·230	408·265	20·206
20	42·956	21·494	460·978	21·470
30	45·821	22·926	524·582	22·904
40	49·097	24·562	602·298	24·542
50	52·876	26·451	698·629	26·432
88°	57·284	28·654	820·035	28·636
10	62·494	31·258	976·036	31·242
20	68·745	34·382	1181·14	34·368
30	76·386	38·202	1458·36	38·188
40	85 936	42·976	1845·91	42·964
50	98·215	49·114	2411·19	49·104
89°	114·586	57·299	3282·14	57·290
10	137·505	68·757	4726·57	68·750
20	171·883	85·946	7385·65	85·940
30	229·181	114·593	13130·6	114·589
40	343·774	171·888	29544·6	171·885
50	687·549	343·775	118180	343·774
90°	Infinite	Infinite	Infinite	Infinite

LONDON:
PRINTED BY WILLIAM CLOWES AND SONS, LIMITED, STAMFORD STREET
AND CHARING CROSS.